THE SCIENTIFIC ASPECT OF

THE SUPERNATURAL

(1866)

THE
SCIENTIFIC ASPECT
OF THE
SUPERNATURAL:

INDICATING THE DESIRABLENESS OF AN
EXPERIMENTAL ENQUIRY
BY MEN OF SCIENCE INTO THE ALLEGED
POWERS OF
CLAIRVOYANTS AND MEDIUMS.

BY

ALFRED RUSSEL WALLACE.

British Library Cataloguing-in-Publication Data
A catalogue record for this book is available from the
British Library

Alfred Russel Wallace

Alfred Russel Wallace was born on 8th January 1823 in the village of Llanbadoc, in Monmouthshire, Wales.

At the age of five, Wallace's family moved to Hertford where he later enrolled at Hertford Grammar School. He was educated there until financial difficulties forced his family to withdraw him in 1836. He then boarded with his older brother John before becoming an apprentice to his eldest brother, William, a surveyor. He worked for William for six years until the business declined due to difficult economic conditions.

After a brief period of unemployment, he was hired as a master at the Collegiate School in Leicester to teach drawing, map-making, and surveying. During this time he met the entomologist Henry Bates who inspired Wallace to begin collecting insects. He and bates continued exchanging letters after Wallace left teaching to pursue his surveying career. They corresponded on prominent works of the time such as Charles Darwin's *The Voyage of the Beagle* (1839) and Robert Chamber's *Vestiges of the Natural History of Creation* (1844).

Wallace was inspired by the travelling naturalists of the day and decided to begin his exploration career collecting specimens in the Amazon rainforest. He explored the Rio Negra for four years, making notes on the peoples and languages he encountered as well as the geography, flora, and fauna. On his return voyage his ship, Helen, caught fire and he and the crew were stranded for ten days before being picked up by the Jordeson, a brig travelling from Cuba to London. All of his specimens aboard Helen had been lost.

After a brief stay in England he embarked on a journey to the Malay Archipelago (now Singapore, Malaysia, and Indonesia). During this eight year period he collected more than 126,000 specimens, several thousand of which represented new species to science. While travelling, Wallace refined his thoughts about evolution and in 1858 he outlined his theory of natural selection in an article he sent to Charles Darwin. This was published in the same year along with Darwin's own theory. Wallace eventually published an account of his travels *The Malay Archipelago* in 1869, and it became one of the most popular books of scientific exploration in the 19th century.

Upon his return to England, in 1862, Wallace became a staunch defender of Darwin's landmark work *On the Origin of Species* (1859). He wrote responses to those critical of the theory of natural selection, including 'Remarks on the Rev. S. Haughton's Paper on the Bee's Cell, And on the Origin of Species' (1863) and 'Creation by Law' (1867). The former of these was particularly pleasing to Darwin. Wallace also published important papers such as 'The Origin of Human Races and the Antiquity of Man Deduced from the Theory of 'Natural Selection" (1864) and books, including the much cited *Darwinism* (1889).

Wallace made a huge contribution to the natural sciences and he will continue to be remembered as one of the key figures in the development of evolutionary theory.

Wallace died on 7th November 1913 at the age of 90. He is buried in a small cemetery at Broadstone, Dorset, England.

THE SCIENTIFIC ASPECT OF THE SUPERNATURAL (1866)

"The perfect observer in any department of science, will have his eyes as it were opened, that they may be struck at once by any occurrence which *according to received theories ought not to happen*, for these are the facts which serve as clues to new discoveries."--SIR JOHN HERSCHELL. "With regard to the miracle question, I can only say that the word 'impossible' is not to my mind applicable to matters of philosophy. That the possibilities of nature are infinite is an aphorism with which I am wont to worry my friends."--PROFESSOR HUXLEY.

INTRODUCTION.

In the following pages I have brought together a few examples of the evidence for facts usually deemed miraculous or supernatural, and therefore incredible; and I have prefixed to these some general considerations on the nature of miracle, and on the possibility that much which has been discredited as such is not really miraculous in the sense of implying any alteration of the laws of nature. In that sense I would repudiate miracles as entirely as the most thorough sceptic. It may be asked if I have myself seen any of the wonders narrated in the following pages. I answer that I have witnessed facts of a similar nature to some of them, and have satisfied myself of their genuineness; and therefore feel that I have no right to reject the evidence of still more marvellous facts witnessed by others. A single new and strange fact is, on its first announcement, often treated as a miracle, and not believed because it is contrary to the hitherto observed order of nature. Half a dozen such facts, however, constitute a little "order of nature" for themselves. They may not be a whit more understood than at first; but they cease to be regarded as miracles. Thus it will be with the many thousands of facts of which I have culled a few examples here. If but one or two of them are proved to be real, the whole argument against the rest, of "impossibility" and "reversal of the laws of nature," falls to the ground. I would ask any man desirous of knowing the truth, to read the following five works carefully through, and then say whether he can believe that the whole of the *facts* stated in them are to be explained by imposture or self-delusion. And let him remember that if but one or two of them are true, there ceases to be any strong presumption against the truth of the rest. These works are--

1. Reichenbach's Researches on Magnetism, Electricity, Heat, Light, &c., in their relations to the vital force. Translated by Dr. Gregory.
2. Dr. Gregory's Letters on Animal Magnetism.
3. R. Dale Owen's Footfalls on the Boundary of Another World.
4. Hare's Experimental Investigation of the Spirit Manifestations.
5. Home's Incidents of my Life.

All these are easily obtained, except the 4th, which may be had from Mr. J. Burns, 1, Wellington Road, Camberwell, S.

I subjoin a list of the persons whose names I have adduced in the following pages, as having been convinced of the truth and reality of most of these phenomena. I presume it will be admitted that they are *honest* men. If, then, these facts, which many of them declare they have repeatedly witnessed, never took place, I must leave my readers to account for the undoubted *fact* of their belief in them, as best they can. I can only do so by supposing these well known men to have been all fools or madmen, which is to me more difficult than believing they are sane men, capable of observing matters of fact, and of forming a sound judgment as to whether or no they could possibly have been deceived in them. A man of sense will not lightly declare, as many of these do, not only that he has witnessed what others deem absurd and incredible, but that he feels morally certain he was not deceived in what he saw.

LIST.

1. Professor A. De Morgan--Mathematician and Logician.
2. Professor Challis--Astronomer.
3. Professor Wm. Gregory, M.D.--Chemist.
4. Professor Robert Hare, M.D.--Chemist.
5. Professor Herbert Mayo, M.D., F.R.S.--Physiologist.
6. Mr. Rutter--Chemist.
7. Dr. Elliotson--Physiologist.
8. Dr. Haddock--Physician.
9. Dr. Gully--Physician.
10. Judge Edmonds--Lawyer.
11. Lord Lyndhurst--Lawyer.
12. Charles Bray--Philosophical Writer.
13. Archbishop Whately--Clergyman.
14. Rev. W. Kerr, M.A.--Clergyman.
15. Hon. Col. E. B. Wilbraham--Military man.
16. Capt. R. F. Burton--Military man.
17. Nassau E. Senior--Political Economist.
18. W. M. Thackeray--Author.
19. T. A. Trollope--Author.
20. R. D. Owen--Author.
21. W. Howitt--Author.
22. S. C. Hall--Author.

CONTENTS

THE SCIENTIFIC ASPECT
OF THE SUPERNATURAL.

I.--MIRACLES AND MODERN SCIENCE.

A miracle is generally defined to be a violation or suspension of a law of nature, and as the laws of nature are the most complete expression of the accumulated experiences of the human race, Hume was of opinion that no amount of human testimony could prove a miracle. Strauss bases the whole argument of his elaborate work on the same ground, that no amount of testimony coming to us through the depth of eighteen centuries can prove that those laws were ever subverted, which the unanimous experience of men now shows to be invariable. Modern science has placed this argument on a wider basis, by showing the interdependence of all these laws, and by rendering it inconceiveable that force and motion, any more than matter, can be absolutely originated or destroyed. Prof. Tyndall, in his paper on *The*

Constitution of the Universe in the *Fortnightly Review*, says, "A miracle is strictly defined as an invasion of the law of the conservation of energy. To create or annihilate matter would be deemed on all hands a miracle; the creation or annihilation of energy would be equally a miracle to those who understand the principle of conservation." Mr. Lecky in his great work on "Rationalism" shows us that during the last two or three centuries, there has been a continually increasing disposition to adopt secular rather than theological views, in history, politics, and science. The great physical discoveries of the last twenty years have pushed forward this movement with still greater rapidity, and have led to a firm conviction in the minds of most men of education, that the universe is governed by wide and immutable laws, under which all phenomena whatever may be classed, and to which no fact in nature can ever be opposed. If therefore we define miracle as a contravention of any one of these laws, it must be admitted that modern science has no place for it, and we can not be surprised at the many and varied attempts by writers of widely different opinions, to account for or explain away all recorded facts in history or religion, which they believe could only have happened on the supposition of miraculous or supernatural agency. This task has been by no means an easy one. The amount of direct testimony to miracles in all ages is very great. The belief in miracles has been, till very recently, almost universal, and it may safely be asserted that, of those who are, on general grounds, most firmly convinced of the impossibility of events deemed miraculous, few if any have thoroughly and honestly investigated the nature and amount of the evidence that those events really happened. On this subject, however, I do not now intend to enter. It appears to me that the very basis of the whole question has been to some extent misstated and misunderstood, and that there are, in every case of supposed miracle, alternative solutions which remove some of those insuperable difficulties which on the ordinary view undoubtedly attach to them. The first and simplest of these solutions is, when

The apparent miracle may be due to some yet undiscovered law of nature.

Many phenomena of the simplest kind would appear supernatural to men having limited knowledge. Ice and snow might easily be made to appear so to inhabitants of the tropics. The ascent of a balloon would be supernatural to persons who knew nothing of the cause of its upward motion; and we may well conceive that, if no gas lighter than atmospheric air had ever been discovered, and if in the minds of all (philosophers and chemists included), air had become indissolubly connected with the idea of the lightest form of terrestrial matter, the testimony of those who had seen a balloon ascend might be discredited, on the grounds that a law of nature must be suspended, in order that anything could freely ascend through the atmosphere in direct contravention to the law of gravitation.

A century ago, a telegram from 3000 miles' distance, or a photograph taken in five seconds, would not have been believed possible, and would not have been credited on testimony, except by the ignorant and superstitious who believed in miracles. Five centuries ago, the effects produced by the modern telescope and microscope would have been deemed miraculous, and if testified by travellers only as existing in China or Japan, would certainly have been disbelieved. The power of dipping the hand into melted metals unhurt, is a remarkable case of an effect of natural laws appearing to contravene another natural law; and it is one which certainly might have been, and probably has been regarded as a miracle and the fact believed or disbelieved, not according to the amount or quality of the testimony to it, but according to the credulity or supposed superior knowledge of the recipient. About twenty years ago, the fact that surgical operations could be performed on patients in the mesmeric trance without their being conscious of pain, was strenuously denied by most scientific and medical men in this country, and the patients, and sometimes the operators, denounced as impostors; the asserted phenomenon was believed to be contrary to the laws of nature. Now, probably every man of intelligence believes the facts, and it is seen that there must be some as yet unknown law of which they are a consequence. When Castellet informed Réaumur that he had reared perfect silkworms from the eggs laid by a virgin moth, the answer was *Ex nihilo nihil fit*, and the fact was disbelieved. It was contrary to one of the widest and best established laws of nature; yet it is now universally admitted to be true, and the supposed law ceases to be universal. These few illustrations will enable us to understand how some reputed miracles may have been due to yet unknown laws of nature. We know so little of what nerve or life-force really is, how it acts or can act, and in what degree it is capable of transmission from one human being to another, that it would be indeed rash to affirm that under no exceptional conditions could phenomena, such as the apparently miraculous cure of many diseases, or perception through other channels than the ordinary senses, ever take place.

But, it will be said, it is only the least important class of miracles that can possibly be explained in this manner. In many cases dead matter is said to have been endowed with force and motion, or to have been suddenly increased immensely in weight and bulk; things altogether non-terrestrial are said to have appeared on earth, and the orderly progress of the great phenomena of nature is affirmed to have been suddenly interrupted. To render any of these things intelligible or possible from the point of view of modern science, we must have recourse to another postulate, which we may state as follows:--

It is possible that intelligent beings may exist, capable of acting on matter, though they themselves are uncognisable directly by our senses.

That intelligent beings may exist around and among us, unperceived during our whole lives, and yet capable under certain conditions of making their presence known by acting on matter, will be inconceivable to some, and will be doubted by many more, but we venture to say, that no man acquainted with the latest discoveries and the highest speculations of modern science, will deny its *possibility*. The difficulty which this conception presents, will be of quite a different nature from that which obstructs our belief in the possibility of miracle, when defined as a contravention of those great natural laws which the whole tendency of modern science declares to be absolute and immutable. The existence of sentient beings uncognisable by our senses, would no more contravene these laws, than did the discovery of the true nature of the Foraminifera, those structureless gelatinous organisms which exhibit so many of the higher phenomena of animal life without any of that differentiation of parts or specialisation of organs which the necessary functions of animal life seem to require. The existence of such preterhuman intelligences if proved, would only add another and more striking illustration than any we have yet received, of how small a portion of the great cosmos our senses give us cognisance. Even such sceptics on the subject of the supernatural as Hume or Strauss, would probably not deny the validity of the conception of such intelligences, or the abstract possibility of their existence. They would perhaps say, "We have no sufficient proof of the fact; the difficulty of conceiving their mode of existence is great; most intelligent men pass their whole lives in total ignorance of any such unseen intelligences; it is amongst the ignorant and superstitious alone that the belief in them prevails. As philosophers we cannot deny the possibility you postulate, but we must have the most clear and satisfactory proof before we can receive it as a fact."

But it may be argued, even if such beings should exist, they could consist only of the most diffused and subtle forms of matter. How then could they act upon ponderable bodies, how produce effects at all comparable to those which constitute so many reputed miracles? These objectors may be reminded, that all the most powerful and universal forces of nature are now referred to minute vibrations of an almost infinitely attenuated form of matter; and that, by the grandest generalisations of modern science, the most varied natural phenomena have been traced back to these recondite forces. Light, heat, electricity, magnetism, and probably vitality and gravitation, are believed to be but "modes of motion" of a space-filling ether; and there is not a single manifestation of force or development of beauty, but is derived from one or other of these. The whole surface of the globe has been modelled and remodelled, mountains have been cut down to plains, and plains have been grooved and furrowed into mountains and valleys, all by the power of etherial heat vibrations set in motion by the sun. Metallic veins and glittering crystals buried deep down under miles of rock and mountain, have been formed by a distinct set of forces developed by vibrations of the same ether. Every green blade and bright blossom that gladdens the surface of the earth, owes its power of growth and life to those vibrations we call heat and light, while in animals and man the powers of that wondrous telegraph whose battery is the brain and whose wires are nerves, are probably due to the manifestation of a yet totally distinct "mode of motion" in the same all- pervading ether. In some cases we are able to perceive the effects of these recondite forces yet more directly. We see a magnet, without contact, or impact of any ponderable matter capable to our imagination of exerting force, yet overcoming gravity and inertia, raising and moving solid bodies. We behold electricity in the form of lightning riving the solid oak, throwing down lofty towers and steeples, or destroying man and beast, sometimes without a wound. And these manifestations of force are produced by a form of matter so impalpable, that only by its effects can it ever be known to us. With such phenomena everywhere around us, we must admit that if intelligences of what we may call an etherial nature do exist, we have no reason to deny them the use of those etherial forces which are the everflowing fountain from which all force, all motion, all life upon the earth originate. Our limited senses and intellects enable us to receive impressions from, and to trace some of the varied manifestations of etherial motion under phases so distinct as light, heat, electricity, and gravity; but no thinker will for a moment assert that there can be no other possible modes of action of this primal element. To a race of blind men, how utterly inconceivable would be the faculty of vision, how absolutely unknowable the very existence of light and its myriad manifestations of form and beauty. Without this one sense, our knowledge of nature and of the universe could not be a thousandth part of what it is. By its absence our very intellect would have been dwarfed, we cannot say to what extent; and we must almost believe that our moral nature could never have been fully developed without it, and that we could hardly have attained to the dignity and supremacy of man. Yet it is possible and even probable that there may be modes of sensation as superior to all ours, as is sight to that of touch and hearing. In the next chapter we shall consider the bearings of this view of the subject on the more recent developments of so-called supernaturalism.

II.--MODERN MIRACLES VIEWED AS NATURAL PHENOMENA.

One very powerful argument against miracles with men of intelligence (and especially with such as are acquainted with the full scope of the revelations of modern science), is derived from the prevalent assumption that, if real, they are the direct acts of the Deity. The nature of these acts is often such, that no cultivated mind can for a moment impute them to an infinite and supreme being. Few if any reputed miracles are at all worthy of a God; and it is the man of science who is best enabled to form a proper conception of the lofty and unapproachable nature of the attributes which must pertain to the supreme mind of the universe. Strange to say, however, he is in most cases illogical enough to consider the difficulties in the way of this assumption as a valid argument against the facts having ever occurred, instead of being merely one against the mode of interpreting them. He even carries this objection further, by the equally unfounded assumption that any beings who could possibly produce the asserted phenomena must be mentally of a high order, and therefore, if the phenomena do not accord with his ideas of the dignity of superior intelligences, he simply denies the facts without examination. Yet many of these objectors admit that the mind of man is not annihilated at death, and that therefore countless millions of beings are constantly passing into another mode of existence, who, unless a miracle of mental transformation takes place, must be very far inferior to himself. Any arguments, therefore, against the reality of phenomena having been produced by preter-human intelligences, on account of the trivial or apparently useless nature of such phenomena, has really no logical bearing whatever upon the question. The assumption that all preter-human intelligences are more intellectual than the average of mankind, is as utterly gratuitous, and as powerless to disprove facts, as that of the opponents of Galileo when they asserted that the planets could not exceed the perfect number, *seven*, and that therefore the satellites of Jupiter did not exist. Let us now return to the consideration of the probable nature and powers of those preter-human intelligences whose possible existence only it is my object at present to maintain.

I have in the first part of this paper given reasons for supposing that there might be, and probably are, other (and perhaps infinitely varied) modes of etherial motion, than those which our senses enable us to recognise. We must therefore admit that there may be and probably are, organisations adapted to receive impressions from them. In the infinite universe there may be infinite possibilities of sensation, each one as distinct from all the rest as sight is from smell or hearing, and as capable of extending the sphere of the possessor's knowledge and the development of his intellect, as would the sense of sight when first added to the other senses we possess. Beings of an etherial order, if such exist, would probably possess some sense or senses of the nature above indicated, giving them increased insight into the constitution of the universe, and proportionately increased intelligence to guide and direct for special ends those new modes of etherial motion, with which they would in that case be able to deal. Their every faculty might be proportionate to the modes of action of the ether. They might have a power of motion as rapid as that of light or the electric current. They might have a power of vision as acute as that of our most powerful telescopes and microscopes. They might have a sense somewhat analogous to the powers of the last triumph of science, the spectroscope, and by it be enabled to perceive instantaneously the intimate constitution of matter under every form, whether in organised beings or in stars and nebulæ. Such existences possessed of such, to us, inconceivable powers, would not be *supernatural*, except in a very limited and incorrect sense of the term. And if those powers were exerted in a manner to be perceived by us, the result would not be a *miracle*, in the sense in which the term is used by Hume or Tyndall. There would be no "violation of a law of nature;" there would be no "invasion of the law of conservation of energy." Neither matter nor force would ever be created or annihilated, even though it might appear so to us. In an infinite universe the great reservoir of matter and force must be infinite, and the fact that an etherial being should be able to exert force, drawn perhaps from the boundless ether, perhaps from the vital energies of human beings, and make its effects visible to us as an apparent "creation," would be no more a real miracle, than is the perpetual raising of millions of tons of water from the ocean, or the perpetual exertion of animal force upon the earth, both of which we have only recently traced, immediately to the sun, and perhaps remotely to other and varied sources lost in the immensity of the universe. All would be still natural. The great laws of nature would still maintain their inviolable supremacy. We should simply have to confess with a modern man of science, that "our five senses are but clumsy instruments to investigate the imponderables," and might see a new and deeper meaning in the oft-quoted but little heeded words of the great poet, when he tells us that "there are more things in heaven and earth than are dreamt of in our philosophy."

It would appear then, if my argument has any weight, that there is nothing self-contradictory, nothing absolutely inconceivable, in the idea of intelligences uncognisable directly by our senses, and yet capable of acting more or less powerfully on matter. There is only to some minds a high improbability, arising from the supposed absence of all proof that there are such beings. Let direct proof be forthcoming, and there seems no reason why the most sceptical philosopher should refuse to accept it. It would be simply a matter to be investigated and tested like any other question of science. The evidence would have to be collected and examined. The results of the enquiries of different observers would have to be compared. The previous character of the observers for knowledge, accuracy, and honesty, would have to be weighed, and some at least of the facts relied on would have to be re-observed. In this manner only could all sources of error be eliminated, and a doctrine of such overwhelming importance be established as truth. I propose now to inquire whether such proof has been given, and whether the evidence is attainable by any one who may wish to investigate the subject in the only manner by which truth can be reached,--by direct observation and experiment.

The first fact capable of proof is this: that during the last 18 years, while physical science has been progressing with rapid strides, and the growing spirit of rationalism has led to a very general questioning of all facts of a supposed miraculous or supernatural character, a continually increasing number of persons maintain their belief in the existence of beings of the nature of those we have hitherto postulated as a bare possibility. All these persons declare that they have received direct and oft-repeated proofs of the existence of such beings. Most of them tell us they have been convinced against all their previous notions and prepossessions. Very many have previously been materialists, not believing in the existence of any intelligences disconnected from a visible, tangible form, nor in the continued existence of the mind of man after death. At the present moment there are at least three millions of persons in the United States of America, who have received to them satisfactory proofs of the existence of invisible intelligences; and in this country there are many thousands who declare the same thing. A large number of these persons continually receive fresh proofs in the privacy of their own homes, and so much interest is felt in the subject that two periodicals are supported in this city, several on the continent, and a very large number in America, which are exclusively devoted to disseminating information relating to the existence of these invisible intelligences and the means of communicating with them. A little enquiry into the literature of the subject, which is already very extensive, reveals the startling fact, that this revival of so-called supernaturalism is not confined to the ignorant or superstitious, or to the lower classes of society. On the contrary, it is rather among the middle and upper classes that the larger proportion of its adherents are to be found; and among those who have declared themselves convinced of the reality of facts such as have been always classed as miracles, are numbers of literary, scientific, and professional men, who always have borne and still continue to bear high characters, are above the imputation either of falsehood or trickery, and have never manifested indications of insanity. Neither is the belief confined to any one religious sect or party. On the contrary, men of all religions and of no religion are alike to be found in the ranks of the believers; and as already stated, many entire sceptics as to there being any super-human intelligences in the universe, have declared that by the force of direct evidence they have been, however unwillingly, compelled to believe that such intelligences do exist.

Here is certainly a phenomenon altogether unique in the history of the human mind. In examining the evidence of similar prodigies during past ages, we have to make much allowance for early education, and the almost universal pre-existing belief in the possibility and frequent occurrence of miracles and supernatural appearances. In the present day it is a notorious fact that among the educated classes, and especially among students of medicine and science, the scepticism on such subjects is almost universal. But what seems the most extraordinary fact of all, and one that would appear to be absolutely inconsistent with any theory of fraud, imposture, or self-delusion, is, that during the eighteen years which have elapsed since the revival of a belief in the supernatural in America, not one single individual has carefully investigated the subject without accepting the reality of the phenomena, and while thousands have been converted *to* the belief, not one adherent has ever been converted back *from* it. While the peculiarly constituted individuals who are the *media* of the phenomena may be counted by thousands, not one has ever exploded the imposture, if imposture it be. And of the few who receive payment for giving up their time to those who wish to witness the manifestations, it is remarkable that no one has yet tried to be first in the market with a full history of the wonderfully ingenious apparatus and extraordinary dexterity that must have been requisite to make dupes of many millions of people, and to establish a new literature and a new religion. They must be very blind not to see that such a work would be a most profitable speculation.

In order that my readers may judge for themselves whether delusion or deception will best account for these facts, or whether we have indeed made a discovery more important and more extraordinary than any that has yet distinguished the nineteenth century, I propose to bring before them a few witnesses, whose evidence it will be well for them to hear before forming a hasty judgment. I shall call chiefly persons connected with science, art, or literature, and whose intelligence and truthfulness in narrating their own observations, are above suspicion; and I would particularly insist, that no objections of a general kind can have any weight against direct evidence to special facts, many of which are of such a nature that there is absolutely no choice between believing that they did occur, or imputing to all who declare they witnessed them, wilful and purposeless falsehood.

III.--OD-FORCE, ANIMAL MAGNETISM, AND CLAIRVOYANCE.

Before proceeding to adduce the evidence of those persons who have witnessed phenomena which, if real, can only be attributed to preter-human intelligences, it will be well to take note of a series of curious observations on human beings, which prove that certain individuals are gifted with unusual powers of perception, sometimes by the ordinary senses leading to the discovery of new forces in nature, sometimes in a manner which no abnormal power of the ordinary senses will account for, but which imply the existence of faculties in the human mind of a nature analogous to those which are generally termed supernatural, and are attributed to the action of unembodied intelligences. It will be seen that we are thus naturally led up to higher phenomena, and are enabled, to some extent, to bridge over the great gulf between the so-called natural and supernatural.

I wish first to call my reader's attention to the researches of Baron Reichenbach, as detailed in Dr. Gregory's translation of his elaborate work. He observed that persons in a peculiar nervous condition experienced well-marked and definite sensations on contact with magnets and crystals, and in total darkness saw luminous emanations from them. He afterwards found that numbers of persons in perfect health and of superior intellect could perceive the same phenomena. As an example, I may mention that among the numerous persons experimented on by Baron Reichenbach were:--

Dr. Endlicher, Professor of Botany and Director of the Botanic Garden of Vienna.

Dr. Nied, a physician at Vienna, in extensive practice, very active and healthy.

M. Wilhelm Hochstetter, son of Professor Hochstetter of Esslingen.

M. Theodore Kotschy, a clergyman, botanist, and well-known traveller in Africa and Persia; a powerful, vigorous, and perfectly healthy man.

Dr. Huss, Professor of Clinical Medicine, Stockholm, and Physician to the King of Sweden.

Dr. Ragsky, Professor of Chemistry in the Medical and Surgical Josephsakademie in Vienna.

M. Constantin Delhez, a French philologist, residing in Vienna.

Mr. Ernest Pauer, Consistorial Councillor, Vienna.

M. Gustav Auschnetz, Artist, Vienna.

Baron von Oberlaender, Forest Superintendent in Moravia.

All these saw the lights and flames on magnets, and described the various details of their comparative size form and colour, their relative magnitude on the positive and negative poles, and their appearance under various conditions, such as combinations of several magnets, images formed by lenses, &c.; and their evidence exactly confirmed the descriptions already given by the "sensitive" patients of a lower class, whose testimony had been objected to, when the observations were first published.

In addition to these, *Dr. Diesing*, Curator in the Imperial Academy of Natural History at Vienna, and the *Chevalier Hubert von Rainer*, Barrister of Klagenfurt, did not see the luminous phenomena, but were highly sensitive to the various sensations excited by magnets and crystals. About fifty other persons in all conditions of life, of all ages, and of both sexes, saw and felt the same phenomena. In an elaborate review of Reichenbach's work in the "British and Foreign Medico-Chirurgical Review," the evidence of these twelve gentlemen, men of position and science, and three of them medical men, is *completely ignored*, and it is again and again asserted that the phenomena are *subjective*. The only particle of argument to support this view is, that a mesmeric patient was *by suggestion* made to see "lights" as well without as with a magnet. It appears to me, that it would be about as reasonable to tell Gordon Cumming or Dr. Livingstone that they had never seen a real lion, because, by suggestion, a score of mesmeric patients can be made to believe they see lions in a lecture room. Unless it can be proved that Reichenbach and these twelve gentlemen, have none of them sense enough to apply simple tests (which, however, the details of the experiments show, were again and again applied), I do not see how the general objections made in the above-mentioned article, that Reichenbach is not a physiologist, and that he did not apply sufficient tests, can have the slightest weight against the mass of evidence he adduces. It is certainly not creditable to modern science, that these elaborate investigations should be rejected without a particle of disproof; and we can only impute it to the distasteful character of some of the higher phenomena produced, and which it is still the fashion of professors of the physical sciences to ignore without examination. I have seen it stated also, that Reichenbach's theory has been disproved by the use of an electro-magnet, and that a patient could not tell whether the current was on or off. But where is the detail of this experiment published, and how often has it been confirmed, and under what conditions? And if true in one case, how does it affect the question, when similar tests *were* applied to Reichenbach's patients; and how does it apply to facts like this, which Reichenbach gives literally by the hundred? "Prof. D. Endlicher saw on the poles of an electro-magnet, flames forty inches high, unsteady, exhibiting a rich play of colours, and ending in a luminous smoke, which rose to the ceiling and illuminated it." (Gregory's Trans. .) The least the deniers of the facts can do, is to request these well-known individuals who gave their evidence to Reichenbach, to repeat the experiments again under exactly similar conditions, as no doubt in the interests of science they would be willing to do. If then, *by suggestion*, they can all be led to describe equally well defined and varied appearances when only sham magnets are used, the odylic flames and other phenomena will have been fairly shown to be very doubtful. But as long as a few negative statements only are made, and the whole body of facts; testified to by men at least equal in scientific attainments to their opponents are left untouched, no unprejudiced individual can fail to acknowledge that the researches of Reichenbach have established the existence of a vast and connected series of new and important natural phenomena. Doctors Gregory and Ashburner in England, state that they have repeated several of Reichenbach's experiments, under test conditions, and have found them quite accurate.

Mr. Rutter, of Brighton, has made, quite independently, a number of curious experiments, which he has detailed in his little work on "Magnetised Currents and the Magnetoscope," and which were witnessed by hundreds of medical and scientific men. He showed that the various metals and other substances, the contact of a male or female hand, or even of a letter written by a male or female, each produced distinct effects on the magnetoscope. And a single drop of water from a glass in which a homœopathic globule had been dissolved, caused a characteristic motion of the instrument when dropped upon the hand of the operator, even when he did not know the substance employed. Dr. King corroborates these experiments, and states that he has seen a decillionth of a grain of silex, and a billionth of a grain of quinine cause motion by means of this apparatus. Every caution was taken in conducting the experiments, which were equally successful when a third party was placed between Mr. R. and the magnetoscope. Magnets and crystals also produced powerful effects, as indicated by Reichenbach. Yet Mr. Rutter's experiments, like Reichenbach's, are ignored by our scientific men, although during several years he offered every facility for their investigation.

The subject of Animal Magnetism is still so much a disputed one among scientific men, and many of its alleged phenomena so closely border on, if they do not actually reach what is classed as supernatural, that I wish to give a few illustrations of the kind of facts by which it is supported. I will first quote the evidence of Dr. William Gregory, late Professor of Chemistry in the University of Edinburgh, who for many years made continued personal investigations into this subject, and has recorded them in his "Letters on Animal Magnetism," published in 1851. The simpler phenomena of what are usually termed "Hypnotism" and "Electro-Biology," are now universally admitted to be real; though it must never be forgotten, that they too had to fight their way through the same denials, accusations, and imputations, that are now made against clairvoyance and phreno-mesmerism. The same men who advocated, tested and established the truth of the more simple facts, claim that they have done the same for the higher phenomena; the same class of scientific and medical men who once denied the former, now deny the latter. Let us see then if the evidence for the one is as good as it was for the other.

Dr. Gregory defines several stages of clairvoyance, sometimes existing in the same, sometimes in different patients. The chief division, however, is into 1. Sympathy or thought-reading, and 2. True clairvoyance. The evidence for the first is so overwhelming, it is to be met with almost everywhere, and is so generally admitted, that I shall not occupy space by giving examples, although it is, I believe, still denied by the more materialistic physiologists.

Dr. Haddock, residing at Bolton, had a very remarkable clairvoyante (E.) under his care. Dr. Gregory says, "After I returned to Edinburgh, I had very frequent communications with Dr. H., and tried many experiments with this remarkable subject, sending specimens of writing, locks of hair, and other objects, the origin of which was perfectly unknown to Dr. H., and in every case, without exception, E. saw and described with accuracy the persons concerned" ().

Sir Walter C. Trevelyan, Bart., received a letter from a lady in London, in which the loss of a gold watch was mentioned.

He sent the letter to Dr. H. to see if E. could trace the watch. She described the lady accurately, and her house and furniture minutely, and described the watch and chain, and described the person who had it, who, she said, was not a habitual thief, and said further that she could tell her handwriting. The lady, to whom these accounts were sent, acknowledged their perfect accuracy, but said, the description of the thief applied to one of her maids whom she did not suspect, so she sent several pieces of handwriting, including that of both her maids. The clairvoyante immediately selected that of the one she had described, and said--"she was thinking of restoring the watch, saying she had found it." Sir W. Trevelyan wrote with this information, but a letter from the lady crossed his, saying, the girl mentioned before by the clairvoyante, *had restored the watch and said she had found it* (.)

Sir W. Trevelyan communicated to Dr. Gregory another experiment he had made. He requested the Secretary of the Geographical Society to send him the writing of several persons abroad, not known to him, and without their names. Three were sent. E. discovered in each case, where they were; in two of them described their persons accurately; described in all three cases, the cities and countries in which they were, so that they could be easily recognised, and told the time by the clocks, which verified the place by difference of longitude (.)

Many other cases, equally well tested, are given in great

detail by Dr. Gregory; and numerous cases are given of tests of what may be called simple direct clairvoyance. For example, persons going to see the phenomena purchase in any shop they please, a few dozens of printed mottoes, enclosed in nutshells. These are placed in a bag, and the clairvoyante takes out a nutshell and reads the motto. The shell is then broken open and examined, and hundreds of mottoes have been thus read correctly. One motto thus read contained ninety-eight words. Numbers of other equally severe test cases, are given by Dr. Gregory, devised and tried by himself and by other well-known persons.

Now, will it be believed, that in the very elaborate article in the "British and Foreign Medico-Chirurgical Review" already referred to, on Dr. Gregory's and other works of an allied nature, *not one single experiment of this kind is mentioned or alluded to*? There is a great deal of general objection to Dr. Gregory's views, because he was a chemist and not specially devoted to physiology (forgetting that Dr. Elliotson and Dr. Mayo who testify to similar facts, were both specially devoted to physiology) and a few quotations of a general nature only are given; so that no reader could imagine that the work criticised was the result of *observation* or *experiment* at all. The case is a complete illustration of judicial blindness. The opponents dare not impute wilful falsehood to Dr. Gregory, Dr. Mayo, Dr. Haddock, Sir Walter Trevelyan, Sir T. Willshire, and other gentlemen who vouch for these facts;

and yet the facts are of such an unmistakable nature, that without imputing wilful falsehood they cannot be explained away. They are therefore silently ignored, or more probably the records of them are never read. The opponents of Galileo refused to look through his telescope, but they could not thereby annihilate the satellites of Jupiter; neither can the silence or contempt of our modern scientific men blind the world any longer to those grand and mysterious phenomena of mind, the investigation of which can alone conduct us to a knowledge of what we really are.

Dr. Herbert Mayo, F.R.S., late Professor of Anatomy and Physiology in King's College, and of Comparative Anatomy in the Royal College of Surgeons, also gives his personal testimony to facts of a similar nature. In his "Letters on the Truths contained in Popular Superstitions" (2nd. Ed.), he says:--"From Boppard, where I was residing in the years 1845-46, I sent to an American gentleman in Paris a lock of hair, which Col. C___, an invalid then under my care, had cut from his own head and wrapped in writing paper from his own writing desk. Col. C___ was unknown even by name to this American gentleman, who had no clue whatever whereby to identify the proprietor of the hair. And all that he did was to place the paper in the hands of a noted Parisian somnambulist. She stated, in the opinion she gave on the case, that Col. C___ had partial palsy of the hips and legs, and that for another complaint he was in the habit of using a surgical instrument. The patient laughed heartily at the idea of the distant somnambulist having so completely realised him."

Dr. Mayo also announces his conversion to a belief in the truth of phrenology and phreno-mesmerism, and Dr. Gregory gives copious details of experiments in which special care has been taken to avoid all the supposed sources of fallacy in phreno-mesmerism; yet although Dr. Mayo's work is included in the criticism already referred to, none of the facts he himself testifies to, nor the latest opinions he puts forward, are so much as once mentioned.

Dr. Joseph Haddock, a physician, resident and practising at Bolton, who has been already mentioned, has published a work entitled "Somnolism and Psycheism," in which he endeavours to classify the facts of mesmerism and clairvoyance, and to account for them on physiological and psychical principles. The work is well worth reading, but my purpose here is to bring forward one or two facts from those which he gives in an appendix to his work. Nothing is more common than for those who deny the reality of clairvoyance to ask contemptuously, "If it is true, why is not use made of it to discover lost property, or to get news from abroad?" To such, I commend the following statement, of which I can only give an abstract.

On Wednesday evening, December the 20th, 1848, Mr. Wood, grocer, of Cheapside, Bolton, had his cash box with its contents stolen from his counting-house. He applied to the police and could get no clue, though he suspected one individual. He then came to Dr. Haddock to see if the girl, Emma, could discover the thief or the property. When put in *rapport* with Emma she was asked about the lost cash box, and after a few moments she began to talk as if to some one not present, described where the box was, what were its contents, how the person took it, where he first hid it; and then described the person, dress, associations of the thief so vividly, that Mr. Wood recognised a person he had *not the least suspected.* Mr. Wood immediately sought out this person, and gave him the option of coming at once to Dr. Haddock's or to the police office. He chose the former, and when he came into the room Emma started back, told him he was a bad man, and had not on the same clothes as when he took the box. He at first denied all knowledge of the robbery, but after a time acknowledged that he had taken it exactly in the manner described by Emma, and it was accordingly recovered.

Now as the names, place, and date of this occurrence are given, and it is narrated by an English physician, it can hardly be denied without first making some enquiry at the place where it is said to have happened. The next instance is of clairvoyance at a much greater distance. A young man had sailed suddenly from Liverpool for New York. His parents immediately remitted him some money by the mail steamer, but they heard, some time afterwards, that he had never applied for it. The mother came twenty miles to Bolton to see if, by Emma's means, she could learn anything of him. After a little time Emma found him, described his appearance correctly, and entered into so many details as to induce his mother to rely upon her statements, and to request Dr. Haddock to make inquiries at intervals of about a fortnight. He did so, and traced the young man by her means to several places, and the information thus acquired was sent to his parents. Shortly after, Dr. Haddock received information from the father that a letter had arrived from his son, and that "it was a most striking confirmation of Emma's testimony from first to last."

We will now pass to the evidence for the facts of what is termed modern Spiritualism.

IV.--THE EVIDENCE OF THE REALITY OF APPARITIONS.

I now propose to give a few instances in which the evidence of the appearance of preter-human or spiritual beings is as good and definite as it is possible for any evidence of any fact to be. For this purpose I shall use some of the remarkable cases collected and investigated by the Hon. Robert Dale Owen, formerly member of Congress and American Minister at Naples. Mr. Owen is the author of works of a varied character; "Essays," "Moral Physiology," "The Policy of Emancipation," and many others. He has been, I believe, throughout his life a consistent and philosophical sceptic, and his writings show him to be well educated, logical, and extremely cautious in accepting evidence.

In 1855, during his official residence at Naples, his attention seems to have been first attracted to the subject of the "supernatural," by witnessing the phenomena occurring in the presence of Mr. Home. He tells us that "sitting in his own well-lighted apartment, in company with three or four friends, all curious observers like himself," a table and lamp weighing ninety-six pounds "rose eight or ten inches from the floor, and remained suspended in the air while one might count six or seven, the hands of all present being laid upon the table."

He then commenced collecting evidence of so-called supernatural phenomena, occurring *unsought for*, and has brought together in his "Footfalls on the Boundary of another World," the best arranged and best authenticated series of facts which have yet been given to the public on this subject. This work is certainly the most philosophical of its kind that has yet appeared, and perhaps, had it been entitled "A Critical Examination into the Evidence of the Supernatural," which it really is, it would have attracted more attention than it appears to have done. I will here give an abstract of two or three of Mr. Owen's cases, as illustrative of their character, and of the careful manner in which they have been authenticated and tested. The first is one which he calls "The Fourteenth of November." ("Footfalls," .)

On the night between the 14th and 15th of November, 1857, the wife of Captain G. Wheatcroft, residing in Cambridge, dreamed that she saw her husband (then in India.) She immediately awoke, and, looking up, she perceived the same figure standing by her bedside. He appeared in his uniform, the hands pressed across the breast, the hair dishevelled, the face very pale. His large dark eyes were fixed full upon her; their expression was that of great excitement, and there was a peculiar contraction of the mouth, habitual to him when agitated. She saw him even to each minute particular of his dress, as distinctly as she had ever done in her life. The figure seemed to bend forward as if in pain, and to make an effort to speak, but there was no sound. It remained visible, the wife thinks, as long as a minute, and then disappeared. She did not sleep again that night. Next morning she related all this to her mother, expressing her belief that Captain W. was either killed or wounded. In due course a telegram was received to the effect that Captain W. had been killed before Lucknow on the 15th of November. The widow informed the Captain's solicitor, Mr. Wilkinson, that she had been quite prepared for the fatal news, but she felt sure there was a mistake of a day in the date of his death. Mr. Wilkinson then obtained a certificate from the War Office, which was as follows:--

9579. No.___, War Office, 30th January, 1858.

"These are to certify that it appears, by the records in this office that Captain G. Wheatcroft, of the 6th Dragoon Guards, was killed in action on the 15th of Nov.," 1857.

(Signed) "B. Hawes."

A remarkable incident now occurred. Mr. Wilkinson was visiting a friend in London, whose wife has all her life had perception of apparitions, while her husband is a "medium." He related to them the vision of the Captain's widow, and described the figure as it appeared to her, when Mrs. N. instantly said, "That must be the very person I saw on the evening we were talking of India." In answer to Mr. Wilkinson's questions, she said they had obtained a communication from him through her husband, and he had said, that he had been killed in India that afternoon by a wound in the breast. It was about nine o'clock in the evening; she did not recollect the date. On further enquiry, she remembered that she had been interrupted by a tradesman, and had paid a bill that evening; and on bringing it for Mr. Wilkinson's inspection, the receipt bore date the *Fourteenth* of November. In March, 1858, the family of Captain Wheatcroft received a letter from Captain G___ C___, dated Lucknow, 19th of December, 1857, in which he said he had been close to Captain W. when he fell, and that it was on the *fourteenth in the afternoon*, and not on the 15th as reported in Sir Colin Campbell's despatches. He was struck by a fragment of a shell in the breast. He was buried at Dilkoosha, and on a wooden cross at the head of his grave are cut the initials G. W., and the date of his death, 14th of November. The War Office corrected their mistake. Mr. Wilkinson obtained another copy of the certificate in April 1859, and found it in the same words as that already given, only that the 14th of November had been substituted for the 15th.

Mr. Owen obtained the whole of these facts, *directly from the parties themselves*. The widow of Captain Wheatcroft examined and corrected his MSS. and showed him a copy of Captain C.'s letter. Mr. Wilkinson did the same, and Mrs. N___ herself related to him the facts which occurred to her. Mrs. N___ had also related the circumstances to

Mr. Howitt before Mr. Owen's investigations, as he certifies in his "History of the Supernatural," vol. 2., . Mr. Owen also states that he has in his possession both the War Office certificates, the first showing the erroneous and the second the corrected date.

Here we have the same apparition appearing to two ladies unknown to and remote from each other, on the same night; the communication obtained through a third person, declaring the time and mode of death; and all coinciding exactly with the events happening many thousand miles away. We presume the *facts* thus attested will not be disputed; and to attribute the whole to "coincidence," must surely be too great a stretch of credulity, even for the most incredulous.

The next case is one of haunting, and is called,

THE OLD KENT MANOR HOUSE ().

In October, 1857, and for several months afterwards, Mrs. R., the wife of a field officer of high rank, was residing in Ramhurst Manor House, near Leigh, in Kent. From her first occupying it, every inmate of the house was more or less disturbed at night, by knocking and sounds as of footsteps, but more especially by voices, which could not be accounted for. Mrs. R.'s brother, a young officer, heard these voices at night and tried every means to discover the source of them in vain. The servants were much frightened. On the second Saturday in October, Miss S., a young lady who had been in the habit of seeing apparitions from her childhood, came to visit Mrs. R., who met her at the railway station. On arriving at the house Miss S. saw on the threshold two figures, apparently an elderly couple, in old-fashioned dress. Not wishing to make her friend uneasy, she said nothing about them at the time. During the next ten days she saw the same figures several times in different parts of the house, always by daylight. They appeared surrounded by an atmosphere of a neutral tint. On the third occasion they spoke to her, and said that they had formerly possessed that house, and that their name was *Children*. They appeared sad and downcast, and said that they had idolised their property, and that it troubled them to know that it had passed away from their family and was now in the hands of strangers. On Mrs. R. asking Miss S. if she had heard or seen anything, she related this to her. Mrs. R. had herself heard the noises and voices continually, but had seen nothing, and after a month had given up all expectation of doing so, when one day, as she had just finished dressing for dinner, in a well-lighted room with a fire in it, and was coming down hastily, having been repeatedly called by her brother who was impatiently waiting for her, she beheld the two figures standing in the doorway dressed just as Miss S. had described them, but above the figure of the lady, written in the dusky atmosphere, in letters of phosphoric light, the words "Dame Children," and some other words intimating that she was "earth bound." At this moment her brother again called out to her that dinner was waiting, and closing her eyes, she rushed through the figures. Inquiries were made by the ladies as to who had lived in the house formerly, and it was only after four months that they found out, through a very old woman, who remembered an old man, who had told her, that he had in his boyhood assisted to keep the hounds for the Children family, who then lived at Ramhurst. All these particulars Mr. Owen received himself from the two ladies, in Dec., 1858. Miss S. had had many conversations with the apparitions, and on Mr. Owen's inquiring for any details they had communicated, she told him that the husband had said his name was *Richard*, and that he had died in 1753. Mr. Owen now determined if possible to ascertain the accuracy of these facts, and after a long search among churchyards and antiquarian clergymen, he was directed to the "Hasted Papers" in the British Museum. From these he ascertained that "*Richard* Children settled himself at Ramhurst," his family having previously resided at a house called "Childrens," in the parish of Tunbridge. It required further research to determine the date. This was found several months later, in an old "History of Kent," by the same "Hasted," published in 1778, where it is stated that "Ramhurst passed by sale to Richard Children, Esq., who resided here, and died possessed of it in 1755, aged eighty-three years." In the "Hasted Papers" it was also stated, that his son did not live at Ramhurst, and that the family seat after Richard's time was Ferox Hall, near Tunbridge. Since

1816 the mansion has been occupied as a farm house, having passed away entirely from the Children family.

However much any one of these incidents might have been scouted as a delusion, what are we to say to the combination of them? A whole household hear distinct and definite noises of persons walking and speaking. Two ladies see the same appearances, at different times, and under circumstances the least favourable for delusion. The name is given to one by voice, to the other by writing; the date of death is communicated. An independent enquirer by much research, finds out that all these facts are true; that the christian name of the only "Children" who occupied and died in the house was *Richard*, and that his death took place in the year given by the apparition, 1753.

Mr. Owen's own full account of this case, and the observations on it should be read, but this imperfect abstract will serve to show that none of the ordinary modes of escaping from the difficulties of a "ghost story" are here applicable.

At page 195 of Mr. Owen's volume we have a most interesting account of disturbances occurring at the parsonage of Cideville, in the department of Seine Inférieure, France, in the winter of 1850-51. The circumstances gave rise to a trial, and the whole of the facts were brought out by the examination of a great number of witnesses. The Marquis de Mirville collected from the legal record all the documents connected with the trial, including the *procès verbal* of the testimony. It is from these official documents Mr. Owen gives his details of the occurrences.

The disturbances commenced from the time when two boys, aged 12 and 14, came to be educated by M. Tinel, the parish priest of Cideville, and continued *two months and a half* until the children were removed from the parsonage. They consisted of knockings, as if with a hammer on the wainscot; scratchings, shakings of the house so that all the furniture rattled; a din as if every one in the house were beating the floor with mallets, the beatings forming tunes when asked, and answering questions by numbers agreed on. Besides these noises there were strange and unaccountable exhibitions of force. The tables and desks moved about without visible cause; the fire irons flew repeatedly into the middle of the room, windows were broken; a hammer was thrown into the middle of the room, and yet fell without noise, as if put down by an invisible hand; persons standing quite alone had their dresses pulled. On the Mayor of Cideville coming to examine into the matter, a table at which he sat with another person, moved away in spite of their endeavours to hold it back, while the children were standing in the middle of the room; and many other facts of a similar nature were observed repeatedly by numerous persons of respectability and position, every one of whom, going with the intention of finding out a trick, were, after deliberate examination, convinced that the phenomena were not produced by any person present. The Marquis de Mirville was himself one of the witnesses.

The interest of this case consists first, in the evidence having been brought out before a legal tribunal; and secondly, in the remarkable resemblance of the phenomena to those which had occurred a short time previously in America, but had not yet become much known in Europe. There is also the closest resemblance to what occurred at Epworth Parsonage in the family of Wesley's father, and which is almost equally well authenticated.[1] Now when in three different countries, phenomena occur of an exactly similar nature, and which are all open to the fullest examination at the time, and when no trick or delusion is in either case found out but every individual of many hundreds who go to see them become convinced of their reality, the fact of the similarity of the occurrences even in many details, is of great weight as indicating a similar *natural* origin. In such cases we cannot fairly accept the general explanation of "imposture," given by those who have not witnessed the phenomena, when none of those who did witness them, could ever detect imposture.

The examples I have quoted, give a very imperfect idea of the variety and interest of Mr. Owen's work, but they will serve to indicate the nature of the evidence he has in every case adduced, and may lead some of my readers to examine the work itself. If they do so they will see, that similar phenomena to those which puzzled our forefathers at

Epworth Parsonage, and at Mr. Mompesson's at Tedworth, have recurred in our own time, and have been subjected to the most searching examination without any discovery of trick or imposture; and they may perhaps be led to conclude that, though often asserted, it is not yet quite proved that "ghosts have been everywhere banished by the introduction of gas."

V.--MODERN SPIRITUALISM: EVIDENCE OF MEN OF SCIENCE.

We have now come to the consideration of what is more especially termed "modern spiritualism," or those phenomena which occur only in the presence, or through the influence, of peculiarly constituted individuals, hence termed "mediums." The evidence is here so abundant, coming from various parts of the world, and from persons differing widely in education, tastes, and religion, that it is difficult to give any notion of its force and bearing by short extracts. I will first adduce that of three men of the highest eminence in their respective departments--Professor De Morgan, Professor Hare, and Judge Edmonds.

AUGUSTUS DE MORGAN, many years Professor of Mathematics, and now also Dean of University College, London, was educated at Cambridge, where he took his degree as 4th wrangler. He studied for the bar, and has been a voluminous writer on mathematics, logic, and biography. He was for eighteen years Secretary to the Royal Astronomical Society, and was a strong advocate for a decimal coinage. In 1863, a work appeared entitled "From Matter to Spirit, the result of ten years' experience in Spirit Manifestations," by C. D., with a preface by A. B. It is very generally known that A. B. is Prof. De Morgan, and C. D. Mrs. De Morgan. The internal evidence of the preface is sufficient to all who know the Professor's style; it has been frequently imputed to him in print without contradiction, and in the *Athenæum* for 1865, in the "Budget of Paradoxes," he notices the work in such a manner as to show that he accepts the imputation of the authorship and still holds the opinions therein expressed.[2] From this preface, which is well worth reading for its vigorous and sarcastic style, I proceed to give a few extracts:--

"I am satisfied from the evidence of my own senses, of *some* of the facts narrated (in the body of the work), of some others I have evidence as good as testimony can give. I am perfectly convinced that I have both seen and heard, in a manner that should make unbelief impossible, things *called* spiritual, which cannot be taken by a rational being to be capable of explanation by imposture, coincidence, or mistake. So far I feel the ground firm under me." (.)

"Ten years ago, Mrs. Hayden, the well-known American medium, came to my house *alone*. The sitting began immediately after her arrival. Eight or nine persons were present, of all ages and of all degrees of belief and unbelief in the whole thing being imposture. The raps began in the usual way. They were to my ear clear, clean, faint sounds such as would be said to *ring* had they lasted. I likened them at the time to the noise which the ends of knitting-needles would make if dropped from a small distance upon a marble slab, and instantly checked by a damper of some kind. . . . Mrs. Hayden was seated at some distance from the table, and her feet were watched. . . . On being asked to put a question to the first spirit, I begged that I might be allowed to put my question mentally--that is without speaking it, or writing it, or pointing it out to myself on an alphabet--and that Mrs. Hayden might hold both arms extended while the answer was in progress. Both demands were instantly granted by a couple of raps. I put the question and desired the answer might be in one word, which I assigned all mentally. I then took the printed alphabet, put a book upright before it, and bending my eyes upon it proceeded to point to the letters in the usual way. The word *chess* was given by a rap at each letter. I had now reasonable certainty of the following alternative: either some *thought-reading* of a character wholly inexplicable, or such superhuman acuteness on the part of Mrs. Hayden that she could detect the letter I wanted by my bearing, though she (seated six feet from the book which hid my alphabet) could see neither my hand nor my eye, nor at what rate I was going through the letters. I was fated to be driven out of the second alternative before the evening was done.

"At a later period of the evening, when another spirit was under examination, I asked him whether he remembered a certain review which was published soon after his death, and whether he could give me the initials of an epithet (which happened to be in five words) therein applied to himself. Consent having been given, I began my way through the alphabet as above; the only difference of circumstances being that a bright table lamp was now between me and the medium. I expected to be brought up, at say, the letter F; and when my pencil passed that letter without any signal, I was surprised, and by the time I came to K, or thereabouts, I paused, intending to announce a failure. But some one called out, 'You have passed it; I heard a rap long ago.' I began again, and distinct raps came first at C. then at D. I was now satisfied that the spirit had failed; but stopping to consider a little more, it flashed into my mind that C. D. were his own initials, and that he had chosen to commence the *clause which contained the epithet*. I then said nothing but 'I see what you are at; pray go on,' and I then got T (for *The*), then the E. I wanted--of which not a word had been said--and then the remaining four initials. I was now satisfied that the contents of my mind had been read which could not have been detected by my method of pointing to the alphabet, even supposing that could have been seen. . . . The things which I have set down were the beginning of a long series of experiences, many as remarkable as what I have given." --"From Matter to Spirit," Preface, pp. xli. xlii.

From the body of the same work I give one short extract:--"The most remarkable instance of *table moving* with a purpose, which ever came under my notice, occurred at the house of a friend, whose family like my own were staying at the seaside. My friend's family consisted of six persons, and a gentleman, now the husband of one of the daughters, joined them, and I was accompanied by a young member of my own family. No paid person was present. A gentleman who had been expressing himself in a very sceptical manner, not only with reference to spirit manifestations, but on the subject of spiritual existence generally, sat on a sofa two or three feet from the dining room table, round which we were placed. After sitting some time we were directed by the rapping to join hands, and stand up round the table *without touching* it. All did so for a quarter of an hour, wondering whether anything would happen, or whether we were hoaxed by the unseen power. Just as one or two of the party talked of sitting down, the old table, which was large enough for eight or ten persons, moved *entirely by itself* as we surrounded and followed it with our hands joined, went towards the gentleman out of the circle, and literally pushed him up to the back of the sofa till he called out 'Hold, enough.'" --"From Matter to Spirit," .

J. W. EDMONDS, commonly called Judge Edmonds, is a man of considerable eminence. He has been elected a member of both branches of the State Legislature of New York, and was for some time President of the Senate. He has been Inspector of Prisons, and made great improvements in the penitentiary system. After passing through various lower offices, he was made a Judge of the Supreme Court of the United States.[3] This is the highest judicial office in the country; he held it for six years, and then resigned, solely on account of the outcry against him on its being known that he had become convinced on the subject of spiritualism. Since then he has resumed his practice at the bar, and was elected to the important office of Recorder of New York, which however he declined to accept.

The Judge was first induced by some friends to visit a medium, and being astonished at what he saw, determined to investigate the matter, and discover and expose what he then believed to be a great imposture. The following are some of his experiences given in his work on "Spirit Manifestations":--

"On the 23rd April, 1851, I was one of a party of nine who sat round a centre table, on which a lamp was burning, and another lamp was burning on the mantle-piece. And then, in plain sight of us all, that table was lifted at least a foot from the floor, and shaken backwards and forwards as easy as I could shake a goblet in my hand. Some of the party tried to stop it by the exercise of their strength, but in vain; *so we all drew back from the table*, and by the light of those two burning lamps we saw the heavy mahogany table suspended in the air."

At the next *seance* a variety of extraordinary phenomena occurred to him. "As I stood in a corner where no one could reach my pocket, I felt a hand thrust into it, and found afterwards that six knots had been tied in my handkerchief. A bass viol was put into my hand, and rested on my foot, and then played upon. My person was repeatedly touched, and a chair pulled from under me. I felt on one of my arms what seemed to be the grip of an iron hand. I felt distinctly the thumb and fingers, the palm of the hand, and the ball of the thumb, and it held me fast by a power which I struggled to escape from in vain. With my other hand I felt all round where the pressure was, and satisfied myself that it was no earthly hand that was thus holding me fast, nor indeed could it be, for I was as powerless in that grip as a fly would be in the grasp of my hand. It continued with me till I thoroughly felt how powerless I was, and had tried every means to get rid of it." Again, as instances of the intelligence and knowledge of the unseen power, he says that during his journey to Central America, his friends in New York were almost daily informed of his condition. On returning, he compared his own journal with their notes, and found that they had accurately known the day he landed, days on which he was unwell or well, and on one occasion it was said he had a headache, and at the very hour he was confined to his bed by a sick headache 2000 miles away. As another example he says, "My daughter had gone with her little son to visit some relatives 400 miles from New York. During her absence, about four o'clock in the morning, I was told through this spiritual intercourse that the little fellow was very sick. I went after him, and found that at the very hour I received that intelligence, he was very sick, his mother and aunt were sitting up with him and were alarmed for the result." . . . "This will give a general idea of what I was witnessing two or three times a week for more than a year. I was not a believer seeking confirmation of my own notions. I was struggling against conviction. I have not stopped to detail the precautions which I took to guard against deception, self or otherwise. Suffice it to say that in that respect I omitted nothing which my ingenuity could devise. There was no cavil too captious for me to resort to, no scrutiny too rigid or impertinent for me to institute, no inquiry too intrusive for me to make."

In a letter published in the *New York Herald*, August 6, 1853, after giving an abstract of his investigations, he says--"I went into the investigation originally thinking it a deception, and intending to make public my exposure of it. Having, from my researches, come to a different conclusion, I feel that the obligation to make known the result is just as strong. Therefore it is, mainly, that I give the result to the world. I say mainly, because there is another consideration which influences me, and that is, the desire to extend to others a knowledge, which I am conscious cannot but make them happier and better."

I would now ask whether it is possible that Judge Edmonds can have been deceived as to these facts and not be insane. Yet he is still in practice at the bar, and is in the highest repute as a lawyer.

ROBERT HARE, M.D., Emeritus Professor of Chemistry in the University of Pennsylvania, was one of the most eminent scientific men of America. He distinguished himself by a number of important discoveries (among which may be mentioned the Oxy-Hydrogen blowpipe) and was the author of more than 150 papers on scientific subjects, besides others on political and moral questions. In 1853 his attention was first directed to table-turning and allied phenomena, and finding that the explanation of Faraday, which he had at first received as sufficient, would not account for the facts, he set himself to work to devise apparatus which should, as he expected, conclusively prove that no force was exerted but that of the persons at the table. The result was not as he expected, for however he varied his experiments he was in every case only able to obtain results which proved that there *was* a power at work not that of any human being present. But in addition to the *power* there was an *intelligence*, and he was thus compelled to believe that existences not human did communicate with him. It is often asserted by the disbelievers in these phenomena, that no scientific man has fully investigated them. This is not true. No one who has not himself inquired into the facts has a right even to give an opinion on the subject, till he knows what has been done by others in the investigation; and to know this it will be necessary for him to read carefully, among other works, "Hare's Experimental Investigations of the Spirit Manifestations," which has passed through five editions. It is a volume of 460 closely printed 8vo. pages and contains besides the details of his experiments, numerous discussions on philosophical, moral, and theological questions, which manifest great acuteness and logical power. The experiments he made were all through private mediums, and his apparatus was so contrived that the medium could not possibly, under the test condition, either produce the motions, or direct the communications that ensued. For example, the table by its movement caused an index to revolve over an alphabet on a disc, yet even when the medium could not see the disc the index moved to such letters as to spell out intelligent and accurate communications. And when the medium's hands were placed upon a truly plane metal plate, supported on accurately turned metal balls, so that not the slightest impulse could be communicated by her to the table, yet the table still moved easily and intelligently. In another case a medium's hands were suspended in water so as to have no connection with the board on which the water vessel was placed, and yet, at request, a force of 18 lbs. was exerted on the boards as indicated by a spring balance (see pages 40 to 50.) A considerable space is devoted to communications received through the means of the above-named apparatus, describing the future life of human beings, and as far as my own judgment goes, these descriptions, taken as a whole, give us a far more exalted, and at the same time more rational and connected view of spirit life, than do the doctrines of any other religion or philosophy; while they are certainly more conducive to morality, and inculcate most strongly the importance of cultivating to the uttermost every mental faculty with which we are endowed. Even if it be possible to prove that the supposed superhuman source of these communications is a delusion, I would still maintain, that standing on their own merits they give us the best, the highest, the most rational, and the most acceptable ideas of a future state, and must prove the best incentive to intellectual and moral advancement; and I would call upon every thinker to examine the work on this account alone, before deciding against it.

I shall next adduce, very briefly, the testimony of a number of well-known and intelligent Englishmen, to facts of a similar nature witnessed by themselves.

VI.--EVIDENCE OF LITERARY AND PROFESSIONAL MEN TO THE FACTS OF MODERN SPIRITUALISM.

T. ADOLPHUS TROLLOPE was educated at Oxford, and is the well-known author of numerous works of high excellence in the departments of travels, fiction, biography, and history. In 1855 he wrote a letter to Mr. Rymer, of Ealing, which was published in the *Morning Advertiser*, and is reproduced in "Incidents of my Life," 2nd ed., , in which he shows the inaccuracy and unfairness of Sir David Brewster's account of phenomena occurring in the presence of both, at Mr. Rymer's house, and concludes with these words: "I should not, my dear sir, do all that duty, I think, requires of me, in this case, were I to conclude without stating very solemnly, that after very many opportunities of witnessing and investigating the phenomena caused by, or happening to Mr. Home, I am wholly convinced, that be what may their origin, and cause, and nature, they are not produced by any fraud, machinery, juggling, illusion, or trickery, on his part." Again, in a letter to the *Athenæum*, eight years later (dated Florence, March 21, 1863) he says, "I have been present at very many 'sittings' of Mr. Home in England, many in my own house in Florence, some in the house of a friend in Florence. . . . My testimony then is this: I have seen and felt physical facts, wholly and utterly inexplicable, as I believe, by any known and generally received physical laws. I unhesitatingly reject the theory which considers such facts to be produced by means familiar to the best professors of legerdemain."

An opinion so positive as this, from a man of such eminence, who during eight years has had repeated opportunities of witnessing, examining and reflecting on the phenomena, must surely be held as of far more value than the opposite opinion, so frequently put forward by those who have either not witnessed them at all, or only on one or two occasions.

JAMES M. GULLY, M.D., author of "Neuropathy and Nervousness," "Simple Treatment of Disease," "The Water Cure in Chronic Diseases." Of the last work the *Athenæum* said: "Dr. Gully's book is evidently written by a well-educated medical man. This work is by far the most scientific that we have seen on Hydropathy." Dr. Gully was one of the persons present at the celebrated *séance* described in the *Cornhill Magazine* in 1860, under the title "Stranger than Fiction," and he wrote a letter to the *Morning Star* newspaper, confirming the entire truthfulness of that article. He says: "I can state with the greatest positiveness that the record made in the article 'Stranger than Fiction' is in every particular correct; that the phenomena therein related actually took place in the evening meeting; and moreover, that no trick, machinery, sleight-of-hand, or other artistic contrivance, produced what we heard and beheld. I am quite as convinced of this last as I am of the facts themselves." He then goes on to show the absurdity of all suggested explanations of such phenomena as Mr. Home's floating across the room, which he both saw and felt, and the playing of the accordion in several persons' hands, often three yards distance from Mr. Home. But the most important fact is, that Dr. Gully is now one of Mr. Home's most esteemed friends. He receives Mr. Home frequently in his house, and has had ample opportunities of testing the phenomena in private, and of certainly detecting the gigantic and complicated system of deception, if it be such. To most minds this will be stronger proof of the reality of the phenomena, than any facts observed at a single *séance*, or than any unsupported assertion that the thing is impossible.

WILLIAM HOWITT, the well-known author of "Rural Life in England," a variety of historical works exhibiting great research, many excellent works of fiction, and recently a "History of Discovery in Australia," has had extensive opportunities of investigating the phenomena, and can hardly be supposed to be incapable of judging of such palpable facts as these: "Mrs. Howitt had a sprig of geranium handed to her by an invisible hand, which we have planted and it is growing; so that it is no delusion, no fairy money turned into dross or leaves. I saw a spirit hand as distinctly as I ever saw my own. I touched one several times, once when it was handing me a flower." . . . "A few evenings afterwards a lady desiring that the 'Last Rose of Summer' might be played by a spirit on the accordion, the wish was complied with, but in so wretched a style that the company begged that it might be discontinued. This was done, but soon after, evidently by another spirit, the accordion was carried and suspended over the lady's head, and there without any visible support or action on the instrument, the air was played through most admirably, in the view and hearing of all."--Letter from William Howitt to Mr. Barkas, of Newcastle, reprinted in Home's "Incidents of my Life," 2nd. ed., .

Here the fact of the spectators not receiving bad music for good, because they believed it to proceed from a superhuman source, is decidedly in favour of their coolness and judgment; and the fact was one which the senses of ordinary mortals are quite capable of verifying.

The HON. COLONEL WILBRAHAM sent the following letter to Mr. Home. I extract it from the *Spiritual Magazine*:--

"46, Brook Street, April 14th, 1863.

"My dear Mr. Home,--I have much pleasure in stating that I have attended several *séances*, in your presence, at the houses of two of my intimate friends and at my own, when I have witnessed phenomena similar to some of those described in your book, which I feel certain could not have been produced by any trick or collusion whatever. The rooms in which they occurred were always perfectly lighted; and it was impossible for me to disbelieve the evidence of my own senses.--Believe me, yours very truly, E. B. WILBRAHAM."

S. C. HALL, F.S.A., Barrister-at-Law, Editor of the *Art Journal*, and well known in literary, artistic, and philanthropic circles, has written the following letter to the Editor of the *Spiritual Magazine*, 1863, --

"Sir,--I follow the example of Colonel Wilbraham, and desire to record my belief in the statements put forth by Mr. D. D. Home ('Incidents of my Life'). I have myself seen nearly all the marvels he relates; some in his presence, some with other mediums, and some when there was no medium aid (when Mrs. Hall and I sat alone). Not long ago, I must have confessed to disbelief in all miracles; I have seen so many that my faith as a Christian is now not merely outward profession, but entire and solemn conviction. For this incalculable good I am indebted to 'Spiritualism;' and it is my bounden duty to induce knowledge of its power to teach and to make happy. That duty may, for the present, be limited to a declaration of confidence in Mr. Home.--Yours, &c., S. C. HALL."

NASSAU WILLIAM SENIOR, late Master in Chancery, and twice Professor of Political Economy in the University of Oxford, was one, whom it will astonish many persons to hear, had become convinced of the truth and reality of what they in their superior knowledge suppose to be a gross delusion. The following statement is made in the *Spiritual Magazine*, 1864, , which can be, no doubt, authoritatively denied if correct:--"We have only to add, as a further tribute to the attainments and honours of Mr. Senior, that he was by long inquiry and experience, a firm believer in Spiritual power and manifestations. Mr. Home was his frequent guest, and Mr. Senior made no secret of his belief among his friends. He it was, who recommended the publication of Mr. Home's recent work by Messrs. Longmans, and he authorised the publication, under initials, of one of the striking incidents there given, which happened to a near and dear member of his family."

The REV. WILLIAM KERR, M.A., Incumbent of Tipton, in his recent work on "Future Punishment, Immortality, and Modern Spiritualism," thus gives his testimony to the facts:--

"The writer of these pages has, for a length of time, bestowed great attention upon the subject, and is in a position to affirm with all confidence, from his own experience and repeated trials, that the alleged phenomena of Spiritualism are, for by far the most part, the products neither of imposture nor delusion. They are true, and that to the fullest extent. The marvels which he himself has witnessed, in the private retirement of his own home, with only a few select friends, and *without having even so much as ever seen a public medium*, are in many respects fully equal to any of the startling narratives that have appeared in print."

THACKERAY, though a cool-headed man of the world and a close student of human nature, could not resist the evidence of his senses in this matter. Mr. Weld, in his "Last Winter in Rome," , states, that at a dinner shortly after the appearance in the *Cornhill Magazine* of the article entitled "Stranger than Fiction," Mr. Thackeray was reproached with having permitted such a paper to appear. After quietly hearing all that could be said on the subject, Thackeray replied, "It is all very well for you, who have probably never seen any spiritual manifestations, to talk as you do; but had you seen what I have witnessed, you would hold a different opinion." He then proceeded to inform Mr. Weld, and the company, that, when in New York, at a dinner party, he saw the large and heavy dinner table, covered with decanters, glasses, and a complete dessert, rise fully two feet from the ground, the *modus operandi* being, as he alleged, spiritual force. No possible jugglery, he declared, was or could have been employed on the occasion; and he felt so convinced that the motive force was supernatural, that he then and there gave in his adhesion to the truth of Spiritualism, and consequently accepted the article on Mr. Home's *séance*.

The late CHANCELLOR, LORD LYNDHURST, was

another eminent convert to Spiritualism. In the *Spiritual Magazine*, 1863, , it is said, "He was a careful and scrutinising observer of all facts which came under his notice, and had no predilections or prejudices against any, and during the repeated interviews which he has had with Mr. Home, he was entirely satisfied of the nearness of the spiritual world, and of the power of spirits to communicate with those still in the flesh. As to the truth of the mere physical phenomena, he had no difficulty in acknowledging them to the fullest extent; neither did he, like many, make any secret of his conviction, as his friends can testify."

ARCHBISHOP WHATELY was a Spiritualist. Mr. Fitzpatrick in his "Memoirs of Whately" tells us, that the Archbishop had been long a believer in Mesmerism, and latterly in clairvoyance and Spiritualism. "He went from one extreme to another, until he avowed an implicit belief in clairvoyance, induced a lady who possessed it to become an inmate of his house, and some of the last acts of his life were excited attempts at table-turning, and enthusiastic elicitations of spirit-rapping." This converted into plain language means, that the Archbishop examined into the facts before deciding against their possibility; and having satisfied himself by personal experiment of their reality, saw their immense importance, and pursued the investigation with ardour.

DR. ELLIOTSON, who for many years was one of the

most determined opponents of Spiritualism, has at length given way to the irresistible logic of facts. Mr. Coleman thus writes in the *Spiritual Magazine*, 1864, :--"'I am,' Dr. Elliotson said to me, and it is with his sanction that I make the announcement, 'now quite satisfied of the reality of the phenomena. I am not yet prepared to admit that they are produced by the agency of spirits. I do not deny this, as I am unable to satisfactorily account for what I have seen on any other hypothesis. The explanations which have been made to account for the phenomena do not satisfy me, but I desire to reserve my opinion on that point at present. I am free, however, to say, that I regret the opportunity was not afforded me at an earlier period. What I have seen lately has made a deep impression on my mind, and the recognition of the reality of these manifestations, from whatever cause, is tending to revolutionise my thoughts and feelings on almost every subject.'"

CAPTAIN BURTON, of Mecca and Salt Lake City, is not a man to be taken in by a "gross deception," yet note what he says about the Davenport Brothers, who are supposed to have been so often exposed. In a letter to Dr. Ferguson, and published by him, Captain Burton states, that he has seen these manifestations under the most favourable circumstances, in private houses, when the spectators were all sceptics, the doors bolted, and the ropes, tape, and musical instruments provided by themselves. He goes on to say: "Mr. W. Fay's coat was removed while he was securely fastened hand and foot, *and a lucifer match was struck at the same instant, showing us the two gentlemen fast bound, and the coat in the air on its way to the other side of the room.* Under precisely similar circumstances, another gentleman's coat was placed upon him." And he concludes thus: "I have spent a great part of my life in Oriental lands, and have seen there many magicians. Lately I have been permitted to see and be present at the performances of Messrs. Anderson and Tolmaque. The latter showed, as they profess, clever conjuring, *but they do not even attempt what the Messrs. Davenport and Fay succeed in doing.* Finally, I have read and listened to every explanation of the Davenport 'tricks' hitherto placed before the English public, and believe me, if anything would make me take that tremendous leap 'from matter to spirit,' *it is the utter and complete unreason of the reasons by which the manifestations are explained.*"

PROFESSOR CHALLIS, the Plumierian Professor of Astronomy at Cambridge, is almost the only person who, as far I know, has stated his belief in some of these phenomena solely from the weight of testimony in favour of them. In a letter to the *Clerical Journal* of June, (?) 1863, he says:--

"But although I have no grounds, from personal observation, for giving credit to the asserted spontaneous movements of tables, I have been unable to resist the large amount of testimony to such facts, which has come from many independent sources, and from a vast number of witnesses. England, France, Germany, the United States of America, with most of the other nations of Christendom, contributed simultaneously their quota of evidence. . . . *In short, the testimony has been so abundant and consentaneous, that either the facts must be admitted to be such as are reported, or the possibility of certifying facts by human testimony must be given up.*"

VII.--THE THEORY OF SPIRITUALISM.

Many of my readers will, no doubt, feel oppressed by the strange and apparently supernatural phenomena here brought before their notice. They will demand, that if indeed they are to accept them as facts, it must be shown that they form a part of the system of the universe, or at least range themselves under some plausible hypothesis.

There is such an hypothesis--old in its fundamental principle, new in many of its details--which links together all these phenomena, as a department of nature hitherto entirely ignored by science and but vaguely speculated on by philosophy; and it does so without in any way conflicting with the most advanced science or the highest philosophy. According to this hypothesis, that, which for want of a better name, we shall term "spirit," is the essential part of all sensitive beings, whose bodies form but the machinery and instruments by means of which they perceive and act upon other beings and on matter. It is "spirit" that alone feels, and perceives, and thinks--that acquires knowledge, and reasons, and aspires--though it can only do so by means of, and in exact proportion to, the organisation it is bound up with. It is the "spirit" of man that is man. Spirit is mind; the brain and nerves are but the magnetic battery and telegraph, by means of which spirit communicates with the outer world.

Though the spirit is in general inseparable from the living body to which it gives animal and intellectual life (for the vegetative functions of the organism could go on without spirit), there not unfrequently occur individuals so constituted that the spirit can wholly or partially quit the body for a time and return to it again. At death it quits the body for ever. The spirit like the body has its laws, and definite limits to its powers. It communicates with spirit easier than with matter, and in most cases can only perceive and act on matter through the medium of embodied spirit. The spirit which has lived and developed its powers clothed with a human body, will, when it leaves that body, still retain its former modes of thought, its former tastes, feelings, and affections. The new state of existence is a natural continuation of the old one. There is no sudden acquisition of new mental proclivities, no revolution of the moral nature. Just what the embodied spirit had made itself, or had become-- *that*, is the disembodied spirit when it begins its life under new conditions. It is the same in character as before, but it has acquired new physical and mental powers, new modes of manifesting the moral sentiments, wider capacity for acquiring physical and spiritual knowledge. The great law of "continuity," so ably shown by Sir William Grove in his recent address to the British Association at Nottingham, to pervade the whole realm of nature, is thus, according to the

Spiritual theory, fully applicable to our passage into, and progress through a more advanced state of existence--a view which should recommend itself to men of science as being in itself probable, and in striking contrast with the doctrines of theologians, which place a wide gulf between the mental and moral nature of man, in his present, and in his future state of existence.

Now this hypothesis, taken as a mere speculation, is as coherent and intelligible as any speculation on such a subject can be. But it claims to be more than a speculation, since it serves to explain and interpret that vast accumulation of facts of which a few examples only have been here given, and to furnish a more intelligible, consistent, and harmonious theory of the future state of man, than either religion or philosophy has yet put forth.

And first; as to the interpretation of facts. In the simplest phenomena of Animal Magnetism, when the muscles, the senses, and the ideas of the patient, are subject to the will of the operator; spirit acts upon spirit, through the intermediation of a peculiar relation between the magnetic or life power of the two organisms; and thus the magnetiser is enabled by his *will*, to create for the patient an ideal world. In the higher phenomenon of "simple clairvoyance," the spirit is to some extent released from the trammels of body, and perceives by some other processes than those of the ordinary senses. In the still higher clairvoyant state termed "mental travelling" the spirit quits the body (still connected with it however by an etherial link), traverses the earth to any distance, communicates with persons in remote countries, if it has any clue by which to distinguish them, and (perhaps through the mediation of their organisations) can perceive and describe events occurring around them.

Under certain conditions disembodied spirit is able to form for itself a visible body out of the emanations from living bodies in a proper magnetic relation to itself; and, under certain still more favourable conditions, this body can be made tangible. Thus all the phenomena of "mediumship" take place. Gravity is overcome by a form of life-magnetism, induced between the spirit and the medium; visible hands or visible bodies are produced, which sometimes write, or draw, or even speak. Thus departed friends come to communicate with those still living, or at the moment of death, the spirit appears visibly, and sometimes, tangibly to the loved ones in a distant land. All these phenomena would take place far more frequently, were the conditions that alone render communication possible, more general, or more cultivated.

It appears then, that all the strange facts, denied by so many because they suppose them "supernatural," may be due to the agency of beings of a like mental nature to ourselves-- who *are*, in fact, ourselves--but one step advanced on the long journey through eternity. The trivial and fantastic nature of the acts of some of these disembodied spirits, is not to be wondered at, when we consider the myriads of trivial and fantastic human beings who are daily becoming spirits, and who retain, for a time at least, their human natures in their new condition. But the *generally* trivial nature of the acts and communications of spirits (admitting them to be such) may be totally denied. If we saw two or three persons making strange gestures in perfect silence, we might probably think they were idiots; but if we found that two of them were deaf and dumb, and the three were conversing in the language of signs, we should become aware that the gesticulations of their bodies were no more intrinsically absurd than the movements of our lips and features during speech. So, if we realise to ourselves the fact, that spirits can in most cases only communicate with us in certain very limited modes, we shall see, that the true "triviality" consists in objecting to any *mode* of mental converse as being trivial or undignified. Then again, as to the matter of the communications, said to be generally "unworthy of a spirit;" the real question is, are they generally such as would have been unworthy of the same spirit when in the body? We should remember too that, in most cases the spirit has first to satisfy the inquirer of its existence, and in many cases, to do so in the face of a strong prejudice against the very possibility of spirit communication, or even of the very existence of spirit. And the undoubted fact that hundreds and thousands of persons have been so convinced by the phenomena they have witnessed in the presence of mediums, shows, that trivial though they may be, these phenomena are well adapted to satisfy many minds, and thus lead them to receive and inquire into the higher phenomena, which they could otherwise never have been induced to examine.

This hypothesis of the existence of spirit, both in man and out of man, and their possible and actual inter-communications, must be judged, exactly in the same way as we judge any other hypothesis--by the nature and variety of the facts it includes and accounts for, and by the absence of any other mode of explaining so wide a range of facts. The truth and reality of the facts however is one thing--the goodness of the hypothesis is another, and to find a flaw in the hypothesis is not to disprove the facts. I maintain that the facts have now been proved, in the only way in which facts are capable of being proved--viz., by the concurrent testimony of honest, impartial, and careful observers. Most of the facts are capable of being tested by any earnest inquirer. They have withstood the ordeal of eighteen years of ridicule, and of rigid scrutiny, during which their adherents have year by year steadily increased, including men of every rank and station, of every class of mind, and of every degree of talent; while not a single individual who has yet devoted himself to a detailed examination of these facts, has denied their reality. These are characteristics of a new truth, not of a delusion or imposture. The facts therefore are proved.

Before proceeding to consider the nature of the doctrine which Spiritualism unfolds, I would wish to say a few words on a recent work by a well known philosophic author, in which the facts of Spiritualism are for the most part admitted, but are accounted for by a different hypothesis from that which I have here briefly explained. Mr. Charles Bray, author of the "Philosophy of Necessity," "Education of the Feelings," &c., has just published a small volume whose title is--"On Force, its mental and moral correlates; and on that which is supposed to underlie all phenomena: with speculations on Spiritualism, and other abnormal conditions of mind." The latter half of the work is entirely devoted to a consideration of the facts of modern spiritualism, and to an attempt to account for them on philosophical principles. Mr. Bray tells us that he has himself witnessed but few of the phenomena, yet enough to satisfy him that they may be true. He seems to rely more on the overwhelming testimony to the facts by men of admitted intelligence, and to the facts themselves being often of such a nature that they cannot be explained away. He has doubtless been led to this less sceptical frame of mind than is usual in philosophic writers, by his acquaintance with cases of clairvoyance, of one of which he states his experience as follows: "*I have heard* a young girl in the mesmeric state, minutely describe all that was seen by a person with whom she was *en rapport*, and in some cases more than was seen or could be seen, such as the initials in a watch which had not been opened, and also describe persons and scenes at a distance, which I afterwards discovered were correctly described, *beyond a possibility of doubt.*" The italics in this sentence are his own.

Judging from the works mentioned in his book, Mr. Bray seems to have but a limited acquaintance with the literature of Spiritualism, which is the more to be regretted as he has so little personal experience of the phenomena, and is therefore hardly in a position to form a satisfactory hypothesis. He considers, however, that he has formed one which "will account for such facts as are genuine," although he admits that he has not made that searching examination which would alone entitle him to decide which facts were genuine, and which were due to fraud or self-delusion. The theory which he propounds is not at all easy to exhibit in a few words. He says that the force which produces the phenomena of Spiritualism "is an emanation from all brains, the medium increasing its density so as to allow others present to come into communion with it, and the intelligence new to every person present, is that of some brain in the distance acting through this source upon the mind of the medium, or others of the circle." . Again, he speaks of "a mental or thought atmosphere the result of cerebration, but devoid of consciousness till it becomes reflected in our own organisations," . It seems to me that this theory labours under the great objection of being unintelligible.

How are we to understand an "emanation from all brains," a "thought atmosphere," producing force and motion, visible and tangible forms, intelligent communications by sounds or motions, and all the other varied phenomena imperfectly sketched in these pages? How does this "unconscious thought atmosphere" form a visible, tangible, force-exerting hand, which can carry flowers, write, or play complete tunes on an instrument? Does it even account for the simpler, yet still marvellous phenomena of clairvoyance? Let us take one of the best authenticated cases observed by Dr. Gregory. Mottoes enclosed in nutshells are purchased at a shop and the clairvoyant reads them accurately. Now we may safely assume, that in this case no human mind knows the particular nutshell, in which each motto is enclosed. How then does the theory of an "emanation from all brains," or that the clairvoyant is through this emanation acted on "by some mind in the distance," explain the reading of these mottoes? If this "emanation" has the power of reading them itself, and communicates them to the clairvoyant, how can we deny it personality, and in what does it differ from that which we term spirit? If the theory of "spirit" is, as Professor De Morgan says, "ponderously difficult," is not this theory of "brain emanation" still more so? I submit, therefore, that Mr. Bray's hypothesis is not tenable, and that nothing but the supposition of personal minds, existing without, as well as with a human body, and capable, under certain conditions only, of acting on us and on matter, is able to account for the whole range of the phenomena. And this supposition has, I maintain, the advantage of being both intelligible and philosophically probable.

It is however very satisfactory to find a writer of Mr. Bray's standing recognising the subject at all, as one which possesses so much truth in it as to require an elaborate theory to account for the phenomena. This alone is a proof of the convincing nature of the evidence for those facts which our men of science neglect to investigate as *à priori* absurd and impossible. The appearance of Mr. Bray's book may perhaps indicate that a change is taking place in public opinion on the subject of clairvoyance and spiritualism, and it will certainly do good service in drawing the attention of thinkers to a class of phenomena which, above all others, seem calculated to lead to the partial solution of the most difficult of all problems--the origin of consciousness, and the nature of mind.

VIII.--THE MORAL TEACHINGS OF SPIRITUALISM.

We have now to consider whether this vast array of phenomena which claims to put us into communication with beings who have passed into another phase of existence, teaches us anything which may make us wiser and better men. I, myself, believe that it does, and shall endeavour, as briefly as possible, to set forth what the doctrines of modern spiritualism really are.

The hypothesis of Spiritualism not only accounts for all the facts (and is the only one that does so), but it is further remarkable as being associated with a theory of a future state of existence, which is the only one yet given to the world that can at all commend itself to the modern philosophical mind. There is a general agreement and tone of harmony, in the mass of facts and communications termed "spiritual," which has led to the growth of a new literature, and to the establishment of a new religion. The main doctrines of this religion are:--that after death man's spirit survives in an etherial body, gifted with new powers, but mentally and morally the same individual as when clothed in flesh. That he commences from that moment a course of apparently endless progression, which is rapid, just in proportion as his mental and moral faculties have been exercised and cultivated while on earth. That his comparative happiness or misery will depend entirely on himself; just in proportion as his higher human faculties have taken part in all his pleasures here, will he find himself contented and happy in a state of existence in which they will have the fullest exercise. While he who has depended more on the body than on the mind for his pleasures, will, when that body is no more, feel a grievous want, and must slowly and painfully develop his intellectual and moral nature till its exercise shall become easy and pleasureable. Neither punishments nor rewards are meted out by an external power, but each one's condition is the natural and inevitable sequence of his condition here. He starts again from the level of moral and intellectual development to which he has raised himself while on earth.

Now here again we have a striking supplement to the doctrines of modern science. The organic world has been carried on to a high state of development, and has been ever kept in harmony with the forces of external nature, by the grand law of "survival of the fittest" acting upon ever varying organisations. In the spiritual world, the law of the "progression of the fittest" takes its place, and carries on in unbroken continuity that development of the human mind which has been commenced here.

The communion of spirit with spirit is said to be by thought-reading and sympathy, and to be perfect between those whose beings are in harmony with each other. Those who differ widely, have little or no power of intercommunion, and thus are constituted "spheres," which are divisions, not merely of space, but of social and moral sympathetic organisation. Spirits of the higher "spheres" can and do communicate with those below; but these latter cannot communicate at will with those above. But there is for all an eternal progress, a progress solely dependent on the power of will in the development of spirit nature. There are no evil spirits but the spirits of bad men, and even the worst are surely if slowly progressing. Life in the higher spheres has beauties and pleasures of which we have no conception. Ideas of beauty and power become realised by the will, and the infinite cosmos becomes a field where the highest developments of intellect may range in the acquisition of boundless knowledge.

It may be thought, perhaps, that I am here giving merely my own ideal of a future state, but it is not so. Every statement I have made is derived from those despised sources, the rapping table, the writing hand, or the entranced speaker. And to show that I have not even done justice, either to the ideas themselves, or to the manner in which they are often conveyed to us, I subjoin a few extracts from the spoken addresses of one of the most gifted "trance-mediums," Miss Emma Hardinge.

In her address on "Hades," she sums up in this passage

her account of our progress through the spheres:--

"Of the nature of those spheres and their inhabitants, we have spoken from the knowledge of the spirits, dwellers still in Hades. Would you receive some immediate definition of your own condition, and learn how *you* shall dwell, and what your garments shall be, what your mansion, scenery, likeness, occupations? Turn your eyes within, and ask what you have learned, and what you have done in this, the school house for the spheres of spirit land. There--there is an aristocracy, and even royal rank and varying degree, but the aristocracy is one of merit, and the royalty of soul. It is only the truly wise who govern, and as the wisest soul is he that is best, as the truest wisdom is the highest love, so the royalty of soul is truth and love. And within the spirit world, all knowledge of this earth, all forms of science, all revelations of art, all mysteries of space, must be understood. The exalted soul that is then fully ready for his departure to a higher state than Hades, must know all that Earth can teach, and have practised all that Heaven requires. The spirit never quits the spheres of earth, until he is fully possessed of all the life and knowledge of this planet and its spheres. And though the progress may be here commenced, and not one jot of what you learn, or think, or strive for here is lost, yet all achievements must be ultimated there, and no soul can wing its flight, to that which you call in view of its perfection, Heaven, till you have passed through Earth and

Hades, and stand ready in your fully completed pilgrimage, to enter on the new and unspeakable glories of the celestial realms beyond."

Could the philosopher or the man of science picture to himself a more perfect ideal of a future state than this? Does it not commend itself to him as what he could wish, if he could, by his wish, form the future for himself? Yet this is the teaching of that which he scouts as an imposture, or a delusion--as the trickery of knaves, or the ravings of madmen--modern Spiritualism. I quote another passage from the same address, and I would ask my readers to compare the modesty of the first paragraph, with the claims of infallibility usually put forward by the teachers of new creeds, or new philosophies:--

"It is true that man is finite and imperfect; hence his utterances are too frequently the dictation of his own narrow perceptions, and his views are limited by his own finite capacity. But as you judge him, so also ye 'shall judge the angels.' Spirits only present you with the testimony of those who have advanced *one step* beyond humanity, and ask for no credence from man without the sanction of man's judgment and reason. Spirits, then, say that their world, is as the soul or spiritual and sublimated essence of this human world of yours; that in locality the spirit world extends around this planet, as all spirit spheres encircle in zones and belts all other planets, earths and bodies in space, until the sphere of each impinges upon the other, and they form in connection one vast and harmonious system of natural and spiritual worlds throughout the universe."

The effects of vice and ungoverned passions are thus depicted:--"Those spirits have engraved themselves with a fatal passion for vice, but, alas! they dwell in a world where there is no means for its gratification. There is the gambler, who has burnt into his soul the fire of the love of gain; he hovers around earth's gamblers, and, as an unseen tempter, seeks to repeat the now lost joys of the fatal game. The sensualist, the man of violence, the cruel and angry spirit; all who have steeped themselves in crime, or painted their souls with those dark stain spots which they vainly think are of the body only--all these are there, no longer able to enact their lives of earthly vice, but retaining on their souls the deadly mark, and the fatal though ungratified desire for habitual sin; and so these imprisoned spirits, chained by their own fell passions in the slavery of hopeless criminal desires, hover round those who attract them as magnets draw the needle, by vicious inclinations similar to their own. But you say, the soul, by tempting others, must thus sink deeper into crime. Ay, but remember that another point of the spiritual doctrine is the universal teaching of eternal progress." And then she goes on to depict in glowing language how these spirits too, in time lose their fierce passions, and learn how to begin the upward path of knowledge and virtue. But I must leave the subject, as I wish to give one extract from the address of the same gifted lady, on the question "What is Spirit?" as an example of the high eloquence and moral beauty with which all her discourses are inspired:--

"Small, and to some of us even insignificant, as seems the witness of the spirit-circle, its phenomenal gleams are lights which reveal in their aggregate, these solemn truths to us. There we behold foregleams of the powers of soul, which so vastly do transcend the laws of matter. That soul's continued existence and triumph over death; our own embodied spirit's power of communication with the invisible world around us, and its various occult forces. Clairvoyance, clairaudience, prophecy, trance, vision, psychometry, and magnetic healing; how grand and wonderful appears the soul, invested even in its earthly prison house, with all these gleams of powers so full of glorious promise of what we shall be, when the prison gates of matter open wide and set the spirit free! Oh! fair young girls, whose forms of supremest loveliness are nature's crowning gems, forget not, when the great Creator's bounteous hand adorned your blooming spring with the radiance of summer flowers, that He shrined within that casket of tinted beauty, a soul whose glory shall survive the decay of all earthly things, and live in weal or woe, as your generation stamps it with beauty or stains it with sinful ugliness, when springs shall no more return, nor summers melt in the vast and changeless evermore. Lift up your eyes from the beautiful dust of to-day, which to-morrow shall be foul in death's corruption, to the ever-living soul which *you*, not *destiny*, must adorn with immortal beauty. Remember you are spirits, and that the hours of your earthly life are only granted you to shape and form those spirits for eternity. Young men, who love to expand the muscles of mind, and wrestle in mental gladiatorial combats for the triumphant crowns of science, what are all these to the eternal conquests to be won in fields of illimitable science in the realms of immortality? Press on through earth as a means, but only to attain to the nobler, higher colleges of the never-dying life, and use mortal aims as instruments to gild your souls with the splendour that never fades, but which yourselves must win here or hereafter, ere you are fit to pass as graduates in the halls of eternal science. To understand that we are spirits, and that we live for immortality, to know and insure its issues; is not this, to Spiritualists, the noblest though last bright page which God has revealed to us? Is not to read and comprehend this page the true mission of modern Spiritualism? All else is but the phenomenal basis of the science which gives us the assurance that spirit lives. This is one great aim and purpose of modern Spiritualism, to know what the spirit is, and what it must do--how best to live, so that it may most surely array itself in the pure white robes of an immortality which is purged of all mortal sin and earthly grossness."

The teachings of Miss Hardinge agree in substance with those of all the more developed mediums, and I would ask whether it is probable that these teachings have been evolved from the conflicting dogmas of a set of impostors? Neither does it seem a more probable solution, that they have been produced "unconsciously" from the minds of self-deluded men and weak women, since it is palpable to every reader that these doctrines are essentially different in every detail, from those taught and believed by any school of philosophers or any sect of Christians.

This is well shown by their opposing statements as to the condition of mankind after death. In the accounts of a future state given by, or through the best mediums, and in the visions of deceased persons by clairvoyants, spirits are uniformly represented in the form of *human* beings, and their occupations as analogous to those of earth. But in most religious descriptions, or pictures of heaven, they are represented as *winged* beings, as resting on, or surrounded by clouds, and their occupations to be playing on golden harps, or perpetual singing prayer and adoration before the throne of God. How is it, if these visions and communications are but the remodelling of pre-existing, or preconceived ideas by a diseased imagination, that the popular notions are never reproduced? How is it, that, whether the medium be man, woman, or child, whether ignorant or educated, whether English, German, or American, there should be one and the same consistent representation of these preterhuman beings, at variance with popular notions of them, but such as strikingly to accord with the modern scientific doctrine of "continuity?" I submit that this little fact is of itself a strong corroborative argument, that there is some objective truth in these communications.

Equally at variance with each other are the popular and the spiritualistic doctrines as regards the Deity. Our modern religious teachers maintain that they know a great deal about God. They define minutely and critically his various attributes; they enter into his motives, his feelings, and his opinions; they explain exactly what he has done, and why he has done it; and they declare that after death we shall be with him, and shall see and know him. In the teaching of the "spirits" there is not a word of all this. They tell us that they commune with higher intelligences than themselves, but of God they really know no more than we do. They say that above these higher intelligences are others higher and higher in apparently endless gradation, but as far as they know, no absolute knowledge of the Deity himself is claimed by any of them. Is it possible, if these "spiritual" communications are but the evolutions of the minds of weak superstitious or deluded human beings, that they should so completely contradict one of the strongest and most cherished beliefs of the superstitious and the religious, and should agree with that highest philosophy (of which most mediums have certainly never heard), which maintains, that we can know nothing of the almighty, the eternal, the infinite, the *absolute* Being, who must necessarily be not only unknown and unknowable, but even *unthinkable* by finite intelligences.

It is often asked, "What has Spiritualism done--what new facts, or what useful information have the supposed spirits ever given to man?" The true answer to this demand probably is, that it is no part of their mission to give knowledge to man which his faculties enable him to acquire for himself, and the very effort to acquire which is part of his education and preparation for the spiritual life. Direct information on matters of fact is however occasionally given, as the records of Spiritualism abundantly show; for example the recent discovery of an inexhaustible supply of pure water in the great city of Chicago (the want of which rendered it notoriously unhealthy) obtained from an artesian well sunk under the guidance of a medium, after it had been pronounced impracticable by men of science. These and all similar facts are however invariably disbelieved *without inquiry*. I prefer therefore to rest the claims of Spiritualism on its moral uses. I would point to the thousands it has convinced of the reality of another world, to the many it has led to devote their lives to works of philanthropy, to the eloquence and the poetry it has given us, and to the grand doctrine of an ever progressive future state which it teaches. Those who will examine its literature will acknowledge these to be facts. Those who will not examine for themselves either the literature or the phenomena of Spiritualism, should at least refrain from passing judgment on a matter of which they are confessedly and wilfully ignorant.

The subject, of which I have here endeavoured to sketch the outlines in a few pages which may perhaps be read when volumes would lie unopened, is far too wide and too important for this mode of treatment to do any justice to it. I have been obliged entirely to leave out all mention of the historical proofs of similar phenomena occurring in unbroken succession from the earliest ages to the present day. I could not allude to the spread of Spiritualism on the continent with its numbers of eminent converts. I could not refer to the numbers of scientific and medical men, who have been convinced of its truth, but have not made public their belief.

In concluding these imperfect illustrations of a subject so generally tabooed by scientific men, I do not expect or wish to make a single convert. All I claim is, to have shown cause for investigation; to have proved that it is not a subject that can any longer be contemptuously sneered at as unworthy of a moment's enquiry. I feel myself so confident of the truth and objective reality of many of the facts here narrated, that I would stake the whole question on the opinion of any man of science desirous of arriving at the truth, if he would only devote two or three hours a week for a few months to an examination of the phenomena, *before pronouncing an opinion*; for, I again repeat, not a single individual that I have heard of, has done this without becoming convinced of the reality of these phenomena. I maintain therefore finally--that whether we consider the vast number and the high character of its converts, the immense accumulation and the authenticity of its facts, or the noble doctrine of a future state which it has elaborated--the so-called supernatural, as developed in the phenomena of animal magnetism, clairvoyance, and modern Spiritualism, is an experimental science which must add greatly to our knowledge of man's true nature and highest interests, and therefore demands an honest and a thorough examination.

Notes Appearing in the Original Work

[1.] In an article entitled "Spirit Rapping a Century Ago," in an early number of the *Fortnightly Review*, an account is given of the disturbances at Epworth Parsonage, the residence of the Wesley family, and it is attempted to account for them by the supposition that they were entirely produced by Hester Wesley, one of John Wesley's sisters; yet the phenomena, even as related by this writer, are such as no human being could possibly have produced, while the moral difficulties of the case are admitted to be quite as great as the physical ones. Every reader of the article must have perceived how lame and impotent is the explanation suggested; and one is almost forced to conclude that the writer did not believe in it himself, so different is the tone of the first part of the article in which he details the facts, from the latter part in which he attempts to account for them. When taken in connection with other similar occurrences narrated by Mr. Owen, all equally well authenticated, and all thoroughly investigated at the time, it will be impossible to receive as an explanation that they were in every case mere childish tricks, since that will not account for more than a minute fraction of the established facts. If we are to reject all the facts this assumption will not explain, it will be much simpler and quite as satisfactory to deny that there are any facts that need explaining.

[2.] The work is now advertised as by *Professor and Mrs. De Morgan.*

*   *   *   *   *

Editor's Note
[3.] *Actually, Edmonds was on the New York Supreme Court only. Wallace corrected this error when he included the essay in his On Miracles and Modern Spiritualism in 1875.*

Printed in Great Britain
by Amazon